DISCLAIMER

This report was prepared as an account of work sponsored by an agency of the United States Government. Neither the United States Government nor any agency thereof, nor any of their employees, makes any warranty, express or implied, or assumes any legal liability or responsibility for the accuracy, completeness, or usefulness of any information, apparatus, product, or process disclosed, or represents that its use would not infringe privately owned rights. Reference therein to any specific commercial product, process, or service by trade name, trademark, manufacturer, or otherwise does not necessarily constitute or imply its endorsement, recommendation, or favoring by the United States Government or any agency thereof. The views and opinions of authors expressed therein do not necessarily state or reflect those of the United States Government or any agency thereof.

Understanding the Benefits of the Smart Grid

Smart Grid Implementation Strategy

DOE/NETL-2010/1413

June 18, 2010

NETL Contact: Keith Dodrill

Integrated Electric Power Systems Division

Office of Systems, Analyses, and Planning

National Energy Technology Laboratory

www.netl.doe.gov

This page is intentionally blank

TABLE OF CONTENTS

Prepared by:

Booz Allen Hamilton

Joe Miller

Horizon Energy Group

Bruce Renz

Renz Consulting, LLC

DOE Contract Number:

DE-FE000400

Acknowledgements

This report was prepared by Booz Allen Hamilton, Inc. (BAH) for the United States Department of Energy's National Energy Technology Laboratory. This work was completed under DOE NETL Contract Number DE-FE000400, and performed under BAH Task 430.04.

The authors wish to acknowledge the excellent guidance, contributions, and cooperation of the NETL staff, particularly:

Steven Bossart, Integrated Electric Power Systems Division Director, NETL

Keith Dodrill, Integrated Electric Power Systems Division, NETL

This page is intentionally left blank

UNDERSTANDING SMART GRID BENEFITS

EXECUTIVE SUMMARY

Since 2005, a great deal has been accomplished to develop and communicate the concepts that define what we call the smart grid today. A number of studies have concluded that, when viewed at a high level, the benefits of a smart grid far outweigh its costs—that is, when all costs and all benefits applicable to all stakeholders are included. But will the smart grid prove beneficial for everyone?

The transition to the smart grid is fundamentally driven by market forces. The smart grid's ability to improve safety and efficiency, make better use of existing assets, enhance reliability and power quality, reduce dependence on imported energy, and minimize costly environmental impacts are all market forces that have substantial economic value. Recognizing these many attributes, the American Recovery and Reinvestment Act of 2009 (ARRA) was designed to provide additional stimulus, to accelerate the smart grid transition, and thereby realize the benefits sooner.

As a result, the momentum for a smart grid is large and growing, but if it is to be sustained, the smart grid's value must become crystal clear to all stakeholders, especially to residential consumers. Residential electricity consumers are the force that drives our economy as well as the strongest political constituency. Therefore it is imperative that they clearly understand the values of a smart grid and that any concerns they may have with its implementation are addressed. Without this understanding and alignment, the U.S. consumer will lack the interest and motivation to support this market-based transition. Rather than a supporter, the consumer could instead become an obstacle to smart grid progress.

There are two reasons to create a national smart grid. First, today's grid needs to be upgraded because it is aging, inadequate, and outdated in many respects—investment is needed to improve its material condition, ensure adequate capacity exists, and enable it to address the 21st-century power supply challenges. Status quo is not an option. Secondly, the benefits of the smart grid are substantial. These benefits will result from improvements in the following six key value areas:

- Reliability — by reducing the cost of interruptions and power quality disturbances and reducing the probability and consequences of widespread blackouts

- Economics — by keeping downward prices on electricity prices, reducing the amount paid by consumers as compared to the "business as usual" (BAU) grid, creating new jobs, and stimulating the U.S. gross domestic product (GDP).

- Efficiency — by reducing the cost to produce, deliver, and consume electricity

- Environmental — by reducing emissions when compared to BAU by enabling a larger penetration of renewables and improving efficiency of generation, delivery, and consumption

- Security — by reducing the probability and consequences of manmade attacks and natural disasters

- Safety — by reducing injuries and loss of life from grid-related events

The value proposition for the smart grid is compelling from a high-level perspective, but is it a good deal when viewed from each individual beneficiary, particularly the end consumer who ultimately foots the bill? Are the values of its benefits, coupled with the associated costs, risks, and other changes/impacts, worth the massive changes the smart grid makes to today's grid?

To achieve alignment in moving forward with a smart grid, answers to the following questions must be clear:

- What's in it for the utilities?
- What's in it for the consumer?
- What's in it for us (society)?

This paper presents the benefits of the smart grid in each of the key value areas and describes how the smart grid generates these benefits. The totality of benefits is then presented from the perspective of each of the beneficiaries to illustrate how compelling the value proposition is for each.

The results suggest that most stakeholders will enjoy some benefits of a smart grid. So far, the utility community is gaining momentum in moving forward given the value proposition they see. On the other hand, the consumers, particularly the residential consumers, remain reluctant because their perceived value proposition is not viewed by them as compelling. More work is needed to communicate the smart grid benefits to the consumer group.

By far, the biggest winner will be society as a whole. The smart grid is expected to provide benefits to society in the following areas:

- Reduced losses to society from power outages and power quality issues
- Improved operating efficiencies of delivery companies and electricity suppliers will reduce their O&M and capital costs, keeping downward pressure on electricity prices for all consumers.
- Improved National Security
- Improved Environmental Conditions
- Improved Economic Growth

When the societal benefits are understood by consumers, their overall value proposition will be compelling and will create the tipping point needed to sustain and even accelerate the transition that is currently underway in the United States. The smart grid can help us ensure that our national prosperity continues throughout this new century.

HOW DOES THE SMART GRID GENERATE BENEFITS?

INTRODUCTION

The Seven Principal Characteristics of the Smart Grid:

- Enables active participation by consumers— Consumer choices and increased interaction with the grid bring tangible benefits to both the grid and the environment, while reducing the cost of delivered electricity.

- Accommodates all generation and storage options— Diverse resources with "plug-and-play" connections multiply the options for electrical generation and storage, including new opportunities for more efficient, cleaner power production.

- Enables new products, services, and markets— The grid's open-access market reveals waste and inefficiency and helps drive them out of the system while offering new consumer choices such as green power products and a new generation of electric vehicles. Reduced transmission congestion also leads to more efficient electricity markets.

- Provides power quality for the digital economy— Digital-grade power quality for those who need it avoids production and productivity losses, especially in digital-device environments.

- Optimizes asset utilization and operates efficiently— Desired functionality at minimum cost guides operations and allows fuller utilization of assets. More targeted and efficient grid-maintenance programs result in fewer equipment failures and safer operations.

- Anticipates and responds to system disturbances (self-heals) — The smart grid will perform continuous self-assessments to detect, analyze, respond to, and as needed, restore grid components or network sections.

- Operates resiliently against attack and natural disaster— The grid deters or withstands physical or cyber attack and improves public safety.

The deployment of technology solutions that achieve these characteristics will improve how the smart grid is planned, designed, operated, and maintained. These improvements—in each of the key value areas presented above—lead to specific benefits that are enjoyed by all.

The following technology solutions are generally considered when a smart grid implementation plan is developed:

- Advanced Metering Infrastructure (AMI)
- Customer Side Systems (CS)
- Demand Response (DR)
- Distribution Management System/Distribution Automation (DMS)
- Transmission Enhancement Applications (TA)
- Asset/System Optimization (AO)
- Distributed Energy Resources (DER)
- Information and Communications Integration (ICT)

The deployment of these technology solutions is expected to create improvements in the six key value areas—reliability, economics, efficiency, environmental, safety and security.

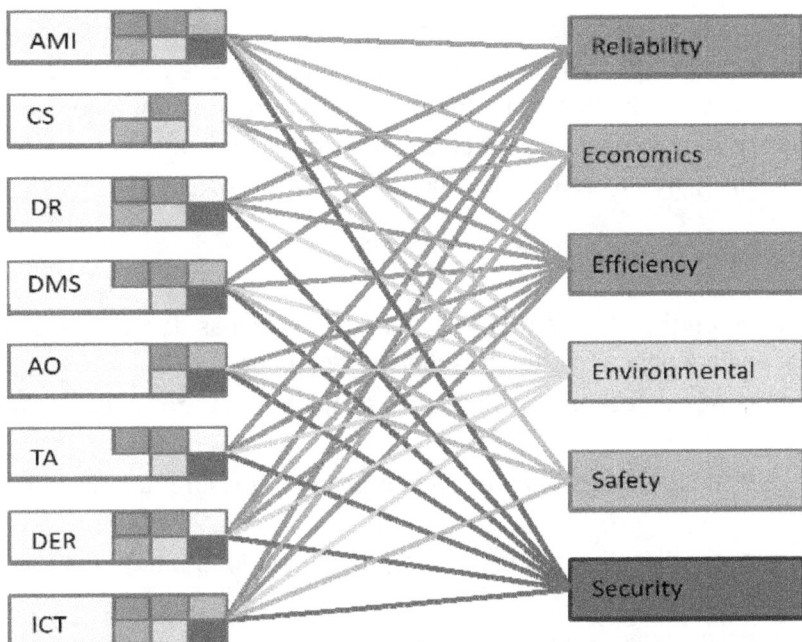

Figure 1: *Technology Solution — Key Value Area Relationships*

Figure 1 identifies the relationships among these technology solutions and the key value areas. This "many-to-many" relationship illustrates the synergy of smart grid solutions, an affect that must be considered when the scope of the smart grid is planned to ensure the benefits are optimized.

The specific benefits that are expected to result from improvements in each of the key value areas are discussed in more detail in the following sections. Achievement of the smart grid vision will depend on how significant these benefits are both collectively and when viewed from the perspective of each individual stakeholder group.

BENEFITS OF IMPROVED RELIABILITY

WHAT DO WE MEAN BY RELIABILITY?

Generally, a reliable grid is one that delivers electricity to consumers when they desire or need it and it is of a quality that supports the consumers' requirements. Improvements in reliability can generally be measured by a reduction in the frequency and duration of outages, a reduction in the number of disturbances due to poor power quality, and virtual elimination of widespread blackouts. Improvements in reliability would create a number of benefits.

Delivery Company Benefits

- Reduced operational costs due to fewer truck rolls, and less demand on call center operations, engineering, and outage response resources
- Improved employee safety as employees are subjected to hazardous conditions less frequently
- Increased revenues as electricity sales are interrupted less frequently and for shorter durations
- Higher customer satisfaction ratings and improved relations with the regulator, the community, etc.
- Reduced capital costs as fewer devices fail in service

The monetized value of these benefits depends on the current state of the delivery company. Some companies will see substantial value while others who may have previously invested in reliability improvements may benefit less. All delivery companies, however, are expected to see some reduction in their costs, and cost reduction keeps downward pressure on future rates—that's good for all consumers.

Electricity Supplier Benefits

- Improved reliability reduces the down time for some generators. When the customer is "off," the generator is selling less of its product. Additionally, a reduction in system transients will reduce the wear and tear on generators and can reduce the time plants are in a forced outage due to system reliability issues.

Residential Consumer Benefits

- Improved level of service with fewer inconveniences caused by outages and poor power quality (resetting electronic devices, no lights, refrigeration, etc.)
- Reduced out-of-pocket costs resulting from loss of sump pumps, spoiled perishables, etc.

Generally, the reliability-related benefits enjoyed by residential consumers are not financially significant. Many never experience financial losses due to reliability problems. But, those who suffer through extended periods of "no power," incurring costs from flooded basements, spoiled food, and other inconveniences might disagree.

Commercial and industrial (C&I) consumers, on the other hand, reap significant benefits from improved reliability. Many of the loads at C&I facilities are electronics based and require a higher level of power quality than ever before. And even short, momentary outages can create havoc. Losses in production and productivity and the impact on worker safety are major. These costs are often passed on in the form of higher prices for their products and services. Therefore, reducing these losses should keep downward pressure on their product prices.

Broader Societal Benefits

- Reduced cost of losses suffered by large consumers from outages. The annual cost of power disturbances to the U.S. economy is enormous (on the order of $100 billion according to EPRI). The smart grid will dramatically reduce this cost, which ultimately helps keep the prices of goods and services lower than they would be otherwise.

- Reduced cost of losses suffered by large consumers from poor power quality. Power quality issues represent a large annual cost to society, estimated to be in the tens of billions of dollars. Merely avoiding the productivity losses of poor quality power to commercial and industrial customers can shed billions of dollars of waste from the economy.

- Virtual elimination of blackouts. The societal cost of a massive blackout is estimated to be in the order of $10 billion per event as described in the 'Final Report on the August 14, 2003 Blackout in the United States and Canada.' The 2003 blackout left over 28 million people in Michigan, New York, Ohio, and other states without electricity for up to four days. The smart grid will be far less vulnerable to such occurrences.

- Improved conditions for economic development. Economic development depends on a reliable, high quality source of electric power. A robust smart grid creates an environment attractive to new investment when compared to one with a poor track record.

HOW THE SMART GRID CREATES IMPROVEMENTS IN RELIABILITY:

The integration of a number of smart grid technologies can contribute to significant improvements in reliability.

- Advanced Metering Infrastructure with communicating smart meters can instantly detect power quality issues and loss of power, enabling system operators to rapidly diagnose system problems and more rapidly restore service.
- Demand Response can reduce the stress on system assets during peak conditions, reducing their probability of failure.
- Distribution Management Systems with ubiquitous sensors, communications, and intelligent controls provide operators with increased situational awareness of the distribution system. Early detection of unhealthy equipment can allow its replacement prior to failure. High-speed, automated line sectionalizing can quickly isolate system problems, thereby limiting the number of customers affected. Intelligent, coordinated control of distributed generation, distributed storage, circuit demand, and adjacent feeders can provide immediate backup when the primary station source is lost.

- Transmission Enhancement Applications employing advanced sensors such as phasor measurement units and new-operator simulation and visualization tools will greatly improve situational awareness. High-speed communications between high-voltage stations and from stations to control centers will enable dynamic management of the transmission system, as compared to today's quasi-steady state control. Automated transmission substations, new digital protection systems, and advanced control devices such as Flexible AC Transmission Systems (FACTS) will combine with this high-speed communications capability to give operators new tools that prevent transmission-level cascading outages.
- Distributed Energy Resources including both local generation and local storage can act as backup sources allowing customers to "ride through" disturbances affecting the normal supply. Advanced microgrid configurations can further integrate the operation of these technologies at the community level.
- Information and Communication Integration, coupled with service oriented IT architecture (SOA), will enable delivery companies to integrate, understand, and act on the vast amount of new information provided by the smart grid to further optimize system reliability.

BENEFITS OF IMPROVED ECONOMICS

WHAT DO WE MEAN BY ECONOMICS?

Improvements in economics are realized when energy and delivery bills paid by consumers are lower than they otherwise would have been. The creation of opportunities for new products and services, the stimulation of economic development and the U.S. GDP, and the creation of new jobs are all elements of improved grid economics.

Delivery Company Benefits

- Numerous opportunities to leverage its resources and enter new markets created by the smart grid such as demand response, microgrid deployment and operations, DER, and others
- Increased revenues as theft of service is reduced, from improved metering accuracy of smart meters over traditional ones, and from shorter power outages
- Improved cash flow from more efficient management of billing and revenue management processes

The monetized value of these benefits depends on the current state of the delivery company. Some will take full advantage of new opportunities, such as those related to the integration of plug-in hybrid vehicles (PHEVs), DER, microgrids, etc. The availability of new information will enable the revenue collection processes to be more efficient and timely. The agility of delivery company organizations will be a key factor in determining how significant the economic benefits will be.

Electricity Supplier Benefits

- New market opportunities for distributed generation and storage will be created.
- The demand for lower cost, environmentally friendly distributed generation and storage will give energy suppliers new options for DER businesses.
- A more robust transmission grid will accommodate larger increases in wind and solar generation.
- A flatter load profile will reduce operating and maintenance (O&M) costs at base-load generating plants. Unfortunately, a flatter load profile will reduce the opportunity for higher-priced peaking units to operate, creating the potential for some stranded assets. One stakeholder's benefit can be another one's liability.

Residential Consumer Benefits

- Downward pressure on energy prices and total customer bills
- Increased capability, opportunity, and motivation to reduce consumption
- Opportunity to interact with the electricity markets through home area network and smart meter connectivity
- Opportunity to reduce transportation costs by using electric vehicles in lieu of conventional vehicles

- Opportunity to sell consumer-produced electricity back to the grid

Recent pricing experiments suggest that residential consumers will change their behavior in response to time-based rates depending on how much those rates vary over time. This reduction in peak and energy consumption will result in reduced electric bills. Generally speaking, these economic benefits may not be major on an individual basis. Commercial and industrial consumers, given their larger demands and higher consumption levels, have a good opportunity to reduce their costs of both energy and demand by participating with the smart grid.

Broader Societal Benefits

- Downward pressure on prices — through improved operating and market efficiencies, reduced supply cost resulting from peak reductions, consumer involvement, and deferral of future capital projects. Overall, markets will be more efficient, resulting in the economically correct prices for electricity, which are expected to be less than they would have been without the capabilities of the smart grid.

- Creation of new jobs — KEMA reported in 2008 that implementation of the smart grid could create 280,000 new jobs and that 140,000 of those jobs would persist following implementation to support its ongoing operation and maintenance.

- Growing the U.S. economy — the demand for new products and services will be created, not only to build the smart grid, but also to support consumers who wish to participate with it. Distributed generation and storage devices, in-home networks, smart appliances, new software applications that support consumer convenience, and energy management features are just a few that come to mind.

- Creation of new electricity markets — enabling society to offer its electricity resources to the market and creating the opportunity to earn a revenue stream on such investments as demand response, distributed generation, and storage.

HOW THE SMART GRID CREATES ECONOMIC IMPROVEMENTS:

The integration of a number of smart grid technologies can contribute to significant improvements in economics.

- Advanced Metering Infrastructure with smart meters and two-way communication capability will provide near-real-time price signals that are linked to wholesale prices to consumers. This information will create the incentive for consumers to respond to prices just as they do for most other products they purchase. This response is expected to reduce the peak demand and the associated prices for electricity.

- Customer Side Systems such as home energy management systems, in-home networks and displays, smart appliances, and others will provide the convenience consumers expect when participating with the smart grid—enabling them to "set it and forget it." Although "simple" from the customer's perspective, these systems will enable complex transactions to take place such as demand response, DER operation, and others.

- Demand Response programs coupled with "smart rate designs" will provide additional incentives to consumers and will create new markets that will stimulate the U.S. economy.

- Distributed Energy Resources including both local generation and local storage can be used to support demand response and in some cases act as resources "for sale" to the electricity market.

- Information and Communication Integration coupled with SOA will integrate the huge volume of new data so that accurate prices signals can be generated, billing information can be collected, and the timely settlement of the vast number of new transactions can be accomplished.

Other technologies such as DMS, TA, and AO that create operational efficiencies will bring costs down over time, leading to a sustained downward pressure on prices. Improved electric system efficiency of the smart grid, also leading to major economic benefits, are discussed in the next section.

BENEFITS OF IMPROVED EFFICIENCY

WHAT DO WE MEAN BY EFFICIENCY?

Efficiency improvements will reduce the cost of producing, delivering and consuming electricity. Reducing the O&M and capital investment costs, as well as the amount of energy used by consumers, will keep downward pressure on future prices and will help the United States more effectively utilize its precious resources.

Delivery Company Benefits

- Increase asset utilization —"getting more through existing assets"
- Reduction in lines losses on both transmission and distribution
- Reduction in transmission congestion costs
- Reductions in peak load and energy consumption leading to deferral of future capital investments
- Increased asset data and intelligence enabling advanced control and improved operator understanding
- Reduction in capital expenditures due to improved utilization of existing assets
- Extended life of system assets through improved asset "health" management
- Improved employee productivity through the use of smart grid information that improves O&M processes
- Improved load forecasting enabling more accurate predictions on when new capital investments are needed
- Reduced use of inefficient generation to meet system peaks

These efficiency benefits should enable delivery companies to keep their costs of service down, resulting in less need to increase rates or at least a delay in how often rates must be increased. Unfortunately, some of these benefits will reduce the volume of sales of kilowatt hours (KWh's) which could impact the companies' total revenues.

Electricity Supplier Benefits

- Reduced transmission congestion gives more competitive generators greater access to markets.
- Efficiency of generation is improved due to flatter load curves.
- Opportunity to expand green power portfolio is due to a more robust transmission grid.
- Fewer forced outages due to a more reliable and efficient transmission system increase unit capacity factors.

Residential Consumer Benefits

- Increased capability, opportunity, and motivation to be more efficient on the consumption end of the value chain
- Increased influence on the electricity market

- Ability to switch from gasoline to electricity for transportation

Conservation and peak load reduction using smart grid technologies give consumers the ability to be more efficient in their consumption. This efficiency improvement helps the delivery company decrease its costs, which can reduce wholesale market prices as demand is reduced. While each consumer's contribution to improving efficiency is small, collectively it can result in significant societal benefits.

Commercial and industrial consumers, given their larger demands and higher consumption levels, can have a more significant impact on efficiency. These larger consumers have already begun participation in demand response programs across the United States to begin working on peak load reduction. Many have also installed combined heat and power (CHP) generation which can substantially increase their overall process efficiency by utilizing the heat from generation rather than rejecting it to the environment.

Broader Societal Benefits

- Deferral of capital investments as future peak loads are reduced and more accurately forecasted through the combined efforts of consumers and delivery companies. The value and need for more efficient baseload generating units to complement the price responsive actions of consumers will increase. Eliminating or deferring large capital investments in centralized generating plants, substations, and transmission and distribution lines, could reduce overall costs by tens of billions of dollars over a 20-year period according to a 2003 Pacific Northwest National Laboratory (PNNL) report.

- Reduced consumption of KWh's through conservation, demand response, and reduced transmission and distribution (T&D) losses. EPRI projects that with a smart grid, the total U.S. electricity consumption could be reduced by 56 to 203 Billion KWh's by 2030 (1.2–4.3%). Besides providing an economic savings to society, this efficiency improvement provides for a better utilization of our U.S. resources.

- Sustained downward pressure on prices as the smart grid enables these efficiency improvements to endure.

HOW THE SMART GRID CREATES EFFICIENCY IMPROVEMENTS:

The integration of a number of smart grid technologies can contribute to significant efficiency improvements.

- Advanced Metering Infrastructure will give consumers the consumption and pricing information they need to enable them to more effectively participate in peak load reduction and energy conservation.

- Customer Side Systems will give consumers the tools they need to conveniently interact with the complexities of the smart grid, enabling them to take actions that reduce their peak load and their overall electricity consumption.

- Demand Response programs will provide additional incentives down to the residential consumer level so that all willing consumers may participate—not just the large commercial and industrial consumers who are viewed as "low-hanging fruit" by today's aggregators.

- Distribution Management Systems equipped with ubiquitous sensors, two-way communications, intelligent controls, advanced control devices, and new visualization tools will give grid operators the tools they need to reduce distribution losses and improve asset utilization.

- Transmission Enhancement Applications will provide the intelligence and control capability needed to support dynamic ratings for transmission lines and other assets, improve transmission asset utilization, and reduce transmission congestion and losses.

- Asset Optimization applications will integrate smart grid data with system planning tools to greatly increase the accuracy of forecasting when new assets are needed to support growing demand. In addition, information from asset health sensors can be collected and integrated with condition-based maintenance programs to improve the overall health and reliability of assets, reduce their out-of-service times and overall maintenance costs, better predict equipment failure, and extend asset life.

- Distributed Energy Resources including both local generation and local storage can be used to support consumer demand response and act as resources "for sale" to the electricity market, thereby contributing to reducing future peak loads below that experienced without a smart grid. Also, power generated locally does not experience the T&D losses of remote generation.

- Information and Communication Integration technologies coupled with SOA will integrate the huge volume of new data with asset management and operational processes to greatly leverage their performance. Conversion of this data to information that can be acted upon by grid operators will also be accomplished with these technologies.

By improving the efficiency of the electric system, positive impacts can be made on the environmental friendliness of the smart grid. The benefits of these environmental improvements are discussed in next section.

BENEFITS OF ENVIRONMENTAL IMPROVEMENTS

WHAT DO WE MEAN BY ENVIRONMENTAL IMPROVEMENTS?

Environmental improvements result in a reduction in emissions and discharges when compared to "business as usual." These improvements go beyond what has historically been considered, now including reductions in CO_2 from generating units as well as reductions in tail pipe emissions due to the expected deployment of smart grid-enabled electric vehicles.

Delivery Company Benefits

- Increased capability to integrate intermittent renewable resources
- Reduction in emissions as a result of more efficient operation, reduced system losses, and energy conservation
- Opportunity to improve environmental leadership image in the area of improving air quality and reducing its carbon footprint
- Increased capability to support the integration of electric-powered vehicles
- Reduction in frequency of transformer fires and oil spills through the use of advanced equipment failure/prevention technologies

Reduced losses enabled by a smart grid will enable delivery companies to reduce the amount of generation (and hence emissions) needed to serve a given load. Similarly, consumers will be equipped and motivated to more effectively conserve energy, again reducing the amount of generation and emission. And with increased knowledge of the smart grid's state, advanced storage, and new control devices, system operators will be able to integrate additional intermittent renewable generation beyond what can be done with today's grid. These new smart grid capabilities will generate significant emission reductions over BAU.

Electricity Supplier Benefits

- New opportunities for renewable generation and storage created by the ability of the smart grid to support increased levels of intermittent resources

Residential Consumer Benefits

- Increased capability, opportunity, and motivation to shift to electric vehicle transportation
- Improved opportunity to optimize energy-consumption behavior resulting in a positive environmental impact
- Increased opportunity to purchase energy from clean resources, further creating a demand for the shift from a carbon-based to a "green economy"

The shift to a "green economy" will only be successful if residential consumers want it. The smart grid can provide the infrastructure to support this shift by giving consumers the ability to participate in the electric vehicle market, reduce energy consumption to reduce emissions, and to give them choices to decide what type of generation source they want as suppliers. While each consumer's contribution to reducing emissions is small, collectively it can result in significant societal benefits.

Commercial and industrial consumers, given their larger demands and higher consumption levels, can have a more significant impact on reducing emissions. For example, the conversion of large roof-top areas to solar generation can reduce their carbon footprints. Other opportunities will undoubtedly emerge as these larger consumers find ways to leverage their resources to address environmental improvement opportunities.

Broader Societal Benefits

- Reduced CO_2 emissions — the smart grid with its ability to support a deep penetration of electric vehicles could reduce emissions by 60–211 million metric tons in 2030 according to EPRI.

- Reduced emissions — through conservation, demand response, and reduced T&D losses, EPRI projects that with a smart grid, the total U.S. electricity consumption could be reduced by 56 to 203 billion KWh's by 2030 (1.2–4.3%). This reduction in energy production provides a corresponding reduction in all types of emissions.

- Improved public health — the cost of health effects can be reduced by 1 to 8 cents for every mile that a conventional vehicle is not driven in an urban area. A deep penetration of PHEVs can reduce the number of miles driven by conventional vehicles (CVs) substantially.

HOW THE SMART GRID CREATES ENVIRONMENTAL IMPROVEMENTS:

The integration of a number of smart grid technologies can contribute to significant environmental improvements.

- Advanced Metering Infrastructure will give consumers the information and control to more effectively manage (reduce or shift) their energy consumption. In addition, consumers will be given new options for selecting how their electricity is generated.

- Customer Side Systems will give consumers the tools they need to conveniently interact with the complexities of the smart grid and to take advantage of their desires to positively impact the environment.

- Demand Response, along with energy storage, will make intermittent renewable resources more viable, thereby increasing their percentages in the national supply portfolio.

- Distribution Management Systems equipped with ubiquitous sensors, two-way communications, advanced control devices, and new visualization tools will give grid operators the tools they need to reduce distribution losses and improve asset utilization, thereby reducing the generation capacity and production needed to serve a given load. Reduced generation translates to reduced emissions of all types. DMS will also allow deep penetration of PHEVs, with all their environmental benefits.

- Transmission Enhancement Applications will provide the intelligence and control capability needed to support the transfer of large blocks of intermittent renewable resources, as well as more efficient conventional generation to the load centers. These applications will improve transmission asset utilization and reduce transmission congestion and losses resulting in fewer emissions of all types.

- Asset Optimization applications will support both DMS and TA to reduce system losses, thereby reducing emissions of all types. Improved equipment failure prediction/prevention will reduce the environmental impact associated with events such as transformer fires and oil spills.

- Distributed Energy Resources will include deep penetration of renewables—perhaps up to 20% of the generation portfolio, new storage technologies, and highly efficient combined heat and power units, all enabled by smart grid technologies. DER in conjunction with large renewable generation farms will be supported by the smart grid, with both contributing to a significant reduction in emissions.

- Information and Communication Integration technologies will support the exchange of vast amounts of information to allow consumers and grid operators to optimize the utilization of these new "green" resources in a reliable and economic fashion.

The smart grid can provide substantial benefits as described in the previous sections. The next section describes the last improvement area, Security and Safety, one that is not often considered by consumers.

IMPROVEMENTS IN SECURITY AND SAFETY

WHAT DO WE MEAN BY SECURITY AND SAFETY?

Improvements in security increase the robustness and resiliency of the grid from a physical perspective and a cyber perspective, thereby reducing the probability and consequences of man-made attacks and natural disasters. In addition, reductions in oil imports made possible by the smart grid enhance national security by increasing U.S. energy independence.

Improvements in safety reduce the hazards inherent in an energized electric system as well as the time of exposure to those hazards.

Delivery Company Benefits

- Reduction in the probability that a deliberate man-made cyber or physical attack could occur and a reduction in the consequences of any that are not detected or prevented
- Improved restoration times following storms and other natural events
- Reduction in theft and vandalism of property due to increased detection capability
- Reduction in injuries and deaths of employees due to reduction in time spent in hazardous situations and the availability of more intelligent systems that support worker safety

These benefits will decrease the probability that extended outages impacting the security of customers will occur by reducing the threat, vulnerability, and consequences of man-made attacks. The increased reliance on digital smart grid technologies will require the deployment of new systems to address cyber security ensuring this critical infrastructure is as "hack proof" as possible. The increased robustness and resilience of the smart grid will also reduce the impact of many natural events that today result in extended outages.

Electric utility work is among the most dangerous of all occupations. The American Public Power Association (APPA) reports that about 1000 fatalities and 7000 flash burns occur annually in the electric utility business. The smart grid with its potential to reduce the frequency and duration of worker exposure to hazardous conditions can reduce these numbers.

Electricity Supplier Benefits

- Reduced exposure of generation plants to potentially damaging and dangerous disturbances due to a more secure transmission system

Residential Consumer Benefits

- Increased peace of mind that the electric grid on which they depend is less likely to be vulnerable to terrorist activity
- Increased ability of grid workers to identify and respond to consumers who depend on electricity for life support when outages or power quality events occur that impact that support

The deployment of large numbers of consumer-owned DER will increase the robustness and resiliency of the grid, providing a natural deterrent to terrorist attack, as well as giving consumers new options when the grid is out of service due to attack or natural disaster. This decentralization, however, will require new techniques that can detect when consumer generation is back-feeding into the system to prevent injury to line workers. The "psychology" of a robust grid will also give consumers a feeling of security in today's uncertain world.

Commercial and industrial consumers would incur large financial impacts as a result of extended outages caused by natural events or terrorist activity. Security improvements created by the smart grid will reduce the probability that outages from these causes will occur.

Broader Societal Benefits

- Increased national security — by reducing U.S. dependence on foreign oil through the use of PHEVs, estimated by PNNL to be up to a 52% reduction. According to Oak Ridge National Laboratory (ORNL), the value of reducing this dependence could be around $13.58 (2004 dollars) for every barrel of oil import reduced.

- Reduction in the probability of widespread and long-term outages due to terrorist activity – The societal cost of a massive blackout is estimated to be in the order of $10 billion per event as described in the "Final Report on the August 14, 2003 Blackout in the United States and Canada." That blackout left over 28 million people in Michigan, New York, Ohio, and other states without electricity for up to four days. Should an outage extend for weeks or even months over a wide section of the United States due to a terrorist attack, the societal cost could be immeasurable.

- Reduction in the number of injuries and deaths associated with the public's contacts with grid assets. The self-healing feature of the smart grid includes the intelligence to ensure the safety of grid workers and the general public. Improved monitoring and decision support systems will quickly identify problems and hazards. The ability to identify equipment that is on the verge of failure is certain to save lives and reduce severe injuries. And reduced outage frequency and duration results in reduced exposure time to health and safety issues.

HOW THE SMART GRID CREATES SECURITY AND SAFETY IMPROVEMENTS:

The integration of a number of smart grid technologies can contribute to improving the security and safety of the electric system.

- Advanced Metering Infrastructure will give grid operators a real-time two-way connection that provides them with the status of individual consumers, including the ability to remotely connect and disconnect their loads.

- Customer Side Systems will give consumers the tools and resources they need to operate autonomously when the grid is distressed.

- Demand Response can be valuable in maintaining electric service when the delivery system is stressed, reducing the probability of an outage or an electrical failure, both of which have health and safety implications.

- Distribution Management Systems equipped with ubiquitous sensors, two-way communications, advanced control devices, and new visualization tools will give grid operators the tools they need to deter, detect, mitigate, respond, and restore from emergency events.

- Transmission Enhancement Applications will provide the intelligence and control capability needed to give grid operators the ability to detect attempts to attack the transmission system, and the capability to respond to prevent cascading impact over a wide area. Advanced protection systems will reduce false trips and more quickly de-energize faulted circuits, both of which can have safety and security implications. In addition, video monitoring of transmission stations will deter terrorist activity.

- Asset Optimization through the deployment of advanced equipment health sensors can identify stressed assets and prevent unexpected failures. Sophisticated monitoring and analysis of grid status will identify marginal operating conditions and allow redistribution of flows to achieve a more secure state. And, security technologies including video cameras and motion detectors can discourage acts of vandalism and sabotage.

- Distributed Energy Resources including both local generation and local storage operating independently, in wide area coordination, and in community microgrid configurations will dramatically reduce the impacts of a natural event or terrorist activities. Their decentralization, diversity in fuel type, and diversity in geographic regions make DER a powerful security tool.

- Information and Communication Integration technologies coupled with SOA will help grid operators better understand the state of the grid following a major event, enabling them to more rapidly and effectively respond to the event.

The smart grid can provide substantial benefits in each of the key value areas. But, are these benefits worth the cost and challenges in making the smart grid a reality?

IS THE SMART GRID REALLY WORTH IT?

WHO ARE THE BENEFICIARIES AND HOW DO THEY BENEFIT?

The fundamental steps to transformation begin with a clear vision of the end goal. Fortunately, the vision for the smart grid has gained clarity over the past few years. The next step is to reach agreement or alignment on that vision. Alignment occurs only after an open and honest debate occurs giving all stakeholders a chance to ask questions, voice concerns, and resolve issues through compromise. Complete alignment on the smart grid vision has not yet occurred, but the level of discussion is high and the interest to resolve differences is sincere. Ultimately, alignment will occur—which is good, but it is not enough.

In order to get real traction, that is, action by the stakeholders to move down the path of smart grid implementation, a motivating force is needed. That force can take two forms—pain or pleasure. Much has been said of the need to modernize the grid (i.e., the pain of doing nothing), but what about the force of pleasure or reward? That force is embodied in the value of the benefits to be enjoyed by each beneficiary group. These benefits define "What's in it for them?"

The potential beneficiaries of the smart grid are wide and varied, but generally speaking, most fit into one or more of the following categories:

- Delivery Companies
- Electricity Suppliers
- Consumers (Residential and C&I)
- Society

Who are these beneficiaries and how do they benefit from the smart grid?

Delivery Companies come in various forms. The state-regulated investor-owned utilities (IOU) have a long history of delivering electricity on a least-cost basis. The IOUs are in business to make a profit and do so by earning a return on the capital investments they make. Additionally, investments made by IOUs that reduce operational and maintenance costs can add to their bottom line, at least until the next rate case. Smart grid investments create the opportunity to realize these savings, provide an opportunity to earn a return on the associated capital investments, and have the potential to improve customer satisfaction. These three opportunities make the smart grid an area of interest to the IOUs.

Electricity is also delivered by non-profit organizations such as cooperatives, municipalities, and public power organizations. They too have an interest in reducing costs to support the desires of their consumer bases to minimize rates.

Given that much of the smart grid investment costs are expected to be recovered through a reduction in operational costs and assuming the delivery companies are able to recover the remaining costs from consumers and earn a return on the investment, it would seem that the delivery companies would be motivated to move forward. This is particularly true if their customers also support (and believe in) the opportunities the smart grid is expected to deliver to them.

The downside for delivery companies is the concern over reduced sales of KWh's. The revenue required by delivery companies to ensure they recover their incremental costs (and return in the case of IOUs) is based on the projected volume of KWh's sold. Solutions to this dilemma, such as the notion of decoupling the revenue from sales, are currently being considered.

The delivery companies should benefit from moving forward with the smart grid, particularly if their risks can be managed.

Electricity Suppliers also come in various forms. Prior to deregulation, electricity was produced by vertically integrated IOUs, and that is still true today in some areas. These IOUs are regulated and are again expected to provide electricity to their customers on a least-cost basis. Their electricity rates include a component to allow them to recover the cost for investing in generating plants plus an allowable rate of return on these investments.

Municipalities and public power entities operating as non-profit organizations also produce or procure electricity on a least-cost basis to ensure the best rates for their customers and citizens. Since deregulation, merchant power producers and new generating companies have formed with the objective of making a profit on the competitive production of electricity.

The smart grid provides new market opportunities for electricity suppliers, as new forms of generation are demanded. But, as the smart grid becomes populated with smaller, more decentralized units and the peak load is flattened as consumers respond to price signals in the new market, the opportunity for peaking units with higher operating cost to operate will diminish.

While new market opportunities exist and the value of baseload generation is expected to increase, the potential for stranded assets, particularly high operating cost peaking units, is real. Generating companies that focus on renewable energy production may find a profitable niche that the smart grid can facilitate.

Consumers are generally broken into three categories—residential (which includes everyone), commercial, and industrial. Over 100 million households exist today in the United States, which is a lot of residential consumers. Given the size of this consumer group, it is very important to ensure that the value of the smart grid is clear and compelling to them. Successfully achieving a smart grid may well depend on how compelling its value proposition is for this group.

WHAT'S IN IT FOR THE RESIDENTIAL CONSUMER?

As individual residential consumers, we are interested in what the smart grid will do for us as individuals. The benefits and costs to the residential consumers include the following:

Benefits

- More reliable service
- Potential bill savings
- Transportation cost savings (PHEVs vs. CVs)
- Information, control, and options for managing electricity more economically and more environmentally friendly

- Option to sell consumer-owned generation and storage resources into the market

Costs

The costs for implementing and operating a smart grid are borne by the consumers.

Are the residential consumers better off with a smart grid or better off under a BAU scenario? How compelling are these benefits? Let's look at two of the specific benefits specified above — potential bill savings and transportation costs savings — to get a general idea of the magnitude of the value proposition for each benefit.

Potential Bill Savings depend on a number of factors, but generally speaking, recent pricing experiments suggest that a range of 10–15% savings in a consumer's electricity bill is realistic. With the average residential electricity bill of $100/month, the savings is $10–$15/ month before considering the increased cost to pay for the new smart grid technologies. For an assumed cost of $5–$10/ month to cover the cost of the new smart grid technologies, the value proposition for the residential consumer reduces to $0–$10/month. This is a positive proposition, but not one that is very compelling.

Transportation Cost Savings (PHEVs vs. CVs)

At current electricity prices, the cost to drive PHEVs using KWh's as fuel is significantly less than the cost for gasoline and diesel fuel. Depending on the miles driven and the relative cost differential between gasoline and diesel fuel, the savings might range between $1,000 to $1,500 per year. The premium paid for the PHEV will obviously impact this savings; however, this value proposition is more compelling—but is it enough to encourage consumers?

These examples suggest that the residential consumer's value proposition, while positive, may not be compelling enough to motivate wide spread acceptance of the smart grid by individual residential consumers. If we are to make a convincing case to them, it may need to be on stronger grounds than cost savings alone.

WHAT'S IN IT FOR THE COMMERCIAL AND INDUSTRIAL CONSUMER?

As C&I consumers, we are interested in what the smart grid will do for our customers and shareholders. The benefits and costs to C&I consumers include the following:

Benefits

- Opportunity to reduce energy and demand charges on bills. The cost of electricity is often a significant portion of the operations budget for these larger users.
- More reliable service resulting in a reduction in the costs of lost production and lost productivity. This reliability ultimately result in lower profits or increased prices for goods and services to their own customers
- There is an option to sell consumer-owned generation and storage resources into the market.

Costs

The costs for implementing and operating a smart grid are borne by these consumers too.

Are the C&I consumers better off with a smart grid or better off under a BAU scenario?

How compelling are these benefits? Generally speaking, large users of electricity should have a greater opportunity to achieve reductions in their energy bills. Also, these users should benefit more from improved reliability since outages and power quality disturbances can create significant costs to their operation when production and productivity are interrupted.

Many of the larger C&I consumers have already invested in interval meters and have implemented special rate designs for energy and demand that enable them to reduce their energy bills. Additionally, many have invested in back-up generating units, uninterruptible power supplies, and redundant power feeds that mitigate the impact of unexpected outages on their operating cost. Despite these actions, the cost of interruptions and power quality issues are estimated to be at least $100 billion annually. These costs mainly impact the larger C&I consumers, resulting in lower profits or higher prices for their goods and services to society.

If the value propositions discussed thus far are not sufficient, how do we make a persuasive argument for the smart grid? Fortunately, the value proposition projected for society is expected to be great—far exceeding all other benefits. Societal benefits are addressed in the next section.

SOCIETAL BENEFITS

SOCIETAL BENEFITS ARE BENEFITS THAT ARE GENERATED THROUGH THE ACTIONS OF SOME INDIVIDUALS AND ENJOYED BY THE REST OF SOCIETY.

Although difficult to monetize, the magnitude of these benefits is great—perhaps great enough to create the tipping point needed to make the smart grid a reality. And since all consumers are also members of society, it is important they understand and value these benefits.

The value proposition for societal benefits answers the question "What's in it for us?"

The smart grid is expected to provide benefits to society in the following areas:

Reduced losses to society from power outages and power quality issues

- Reducing the probability of regional blackouts can prevent significant losses to society. The societal cost of the August 2003 blackout was $8.6 billion.
- Reducing by even 20% the cost of outages and power quality issues, which are estimated to be at least $100 billion annually, would save $20 billion per year.

Improved operating efficiencies of delivery companies and electricity suppliers will reduce their O&M and capital costs, keeping downward pressure on electricity prices for all consumers.

- Reducing T&D Losses which have been estimated at over $25 billion per year by the Business Roundtable by even 10% would save $2.5 billion/year.
- Reducing transmission congestion costs, which range from $4.8 billion to as much as $50 billion annually, by 10%, could save up to $2 billion/year.
- Reduced O&M spending. Significant amounts (at least $40 billion in 2005 excluding fuel and purchased power) are spent annually by IOUs to operate and maintain the power system. While yet to be quantified, even a 10% savings resulting from grid modernization would save $4 billion each year.
- Eliminating or deferring large capital investments in centralized generating plants, substations, and transmission and distribution lines, could reduce overall costs $46–$117 billion dollars over a 20-year period according to a 2003 PNNL report.

Improved National Security

- Reducing the U.S. dependence on foreign oil through the use of PHEVs could be up to 52% based on a recent PNNL report. This is an equivalent of reducing U.S. oil consumption by 6.5 million barrels per day. According to ORNL, the value of reducing this dependence is $13.58 (2004 dollars) for every barrel of oil import reduced, creating a potential societal benefit of over $30 billion/year.
- Reducing the probability (and consequences) of widespread and long-term outages due to terrorist activity could prevent significant societal costs that are immeasurable.

Improved Environmental Conditions

- Reduction in total emissions — Through conservation, demand response, and reduced T&D losses, EPRI projects that with a smart grid, the total U.S. electricity consumption could be reduced by 56 to 203 billion KWh's by 2030 (1.2–4.3%). This reduction in energy production provides a corresponding reduction in all types of emissions.

- Reduction in CO_2 emissions — The smart grid and its ability to support a deep penetration of electric vehicles could reduce emissions by 60–211 million metric tons in 2030 according to EPRI.

- Improved public health — The impact of vehicle particulate emissions in urban areas can be reduced as the number of miles driven by CVs is offset by miles driven by electric vehicles.

- Reduction in the number of injuries and deaths due to contact with grid assets.

Improved Economic Growth

- Creation of new jobs — KEMA reported in 2008 that implementation of the smart grid could create 280,000 new jobs and that 140,000 of those jobs would persist following implementation to support its ongoing operation and maintenance.

- Demand for new products and services — This demand will be created not only to build the smart grid but also to support consumers who wish to participate with it.

- Creation of new electricity markets — Such markets will enable society to offer its electricity resources to the market, creating the opportunity to earn a revenue stream on such investments as demand response, distributed generation, and storage.

- Improved conditions for economic development — Economic development depends on a reliable source of electric power. A robust smart grid creates an environment attractive to new investment when compared to one with a poor track record.

- Reduced wholesale electricity prices compared with BAU – This reduction will be achieved through a reduction in peak loads and energy conservation. Assuming an overall average wholesale price for electricity of approximately $47 per megawatt hour (MWh) and annual sales of approximately 3.5 billion/MWh, a 1% reduction in average price would result in an economic savings to society of over $1.5 billion annually.

- Reduced consumption of KWh's through conservation, demand response, and reduced T&D losses – EPRI projects, that with a smart grid, the total U.S. electricity consumption could be reduced by 56 to 203 billion KWh's by 2030 (1.2–4.3%). Besides providing an economic savings to society, this efficiency improvement provides for a better utilization of our national resources.

- A Synapse Energy Economics study — This study found that increasing energy efficiency, renewable energy, and distributed generation would save an estimated $36 billion annually by 2025.

Further work is needed to quantify these opportunities, but clearly they appear to be quite large. Yet are they large enough to encourage the large residential consumer group to embrace the massive changes required by the smart grid transformation?

CONCLUSIONS

ARE THE BENEFITS OF THE SMART GRID COMPELLING ENOUGH TO UNITE ALL BENEFICIARIES TO MOVE FORWARD?

The benefits enjoyed by each of the beneficiaries have been discussed and the value proposition to most is positive. In summary—

- Delivery companies have the opportunity to significantly reduce their O&M and capital costs and earn a return for their shareholders. The risk of cost recovery is a concern.

- Electricity suppliers have the opportunity to enter new markets created by the smart grid including renewable generation, storage, and DER. Some higher-priced peaking units could be stranded if the smart grid successfully flattens the peak, but flatter peaks also create positive effects for base-load generating plants.

- Consumers have the opportunity to reduce their electricity bills, reduce their losses caused by outages and power quality events, and offer their resources to the market to generate a revenue stream. Consumers bear the incremental cost for implementing the smart grid without a guarantee that the benefits will be realized.

- Society will benefit from a stimulated economy, improved environmental conditions, improved national security, job creation, and a sustained downward pressure on future price increases for electricity. The incremental cost to society is essentially zero.

Past smart grid studies suggest that the value proposition for the smart grid is positive when viewed from an overall perspective.

- According to EPRI, "The grid of the future will require $165 billion over the next 20 years". The benefits to society will be $638 to $802 billion. The benefit to cost ratio is 4 or 5 to 1.

- The West Virginia Smart Grid Implementation Plan identified that costs of $1.9 billion over the next 20 years could yield benefits of $12.6 billion. The benefit to cost ratio is 6.7 to 1.

When viewed from an individual beneficiary group's perspective we see some delivery companies and some electricity suppliers moving ahead, yet the consumers appear tentative and reluctant. Consumers have identified two areas of concern:

The direct financial benefits to consumers are not compelling, particularly when compared to the risks.

As discussed throughout this paper, the individual financial benefits to the largest consumer group, the residential class, although positive, are not large enough to generate widespread enthusiasm.

Consumers who bear the cost of the smart grid are also concerned that the expected benefits may not materialize, leaving them with higher rates and benefits that are "less than advertised."

Other non-financial consumer concerns have not been adequately addressed:

- Privacy — some consumer groups have raised privacy concerns associated with AMI. Not everyone wants their utility to have detailed information on their energy usage patterns, which can also reveal a great deal about their overall lifestyle. How do we protect the privacy of consumers? Other industries have addressed this concern.
- Control and trust — consumer groups have raised questions about whether utilities should be allowed to vary pricing over the course of a day as production costs change. They also worry that smart grid technologies may be just another way for utilities to make extra money from consumers.
- Burdensome new tools — given that many consumers fail to program their video recording systems and thermostats, how likely is it that the average consumer will want to take the time to learn how to use energy management software that can monitor and optimize their energy usage? Can we make it as simple as "set it and forget it"?

Other unanswered questions remain. Clear communications that address the following questions are needed:

- Why do we need to pursue the consumer side (smart meters) before smart grid upgrades are made to the transmission and distribution system?
- Why can't many of the benefits the smart grid provides be done with existing technologies (e.g., existing demand response technologies)?
- All consumers will pay for smart grid investments, but only some will be able to realize all the benefits. Is that fair?
- Will consumers have to purchase additional devices to participate with the smart grid and enjoy its benefits (e.g., home area networks, in-home displays)?
- Will smart grid technologies increase the risk of cyber security breaches resulting in a less secure grid and the leaking of private consumer data?

Engaging consumers in the smart grid transition will require extensive two-way communication and education programs. These programs should address both their financial and non-financial concerns.

- Addressing consumer financial concerns — we must help consumers understand the magnitude of the benefits they will enjoy as individual consumers from a deployed smart grid. Additionally, we must describe the significant societal benefits the smart grid creates for the United States and that those benefits come with little or no additional costs. Together, the consumer and societal value proposition is compelling and both should be included in consumer education programs. Some method of sharing the "risk and reward" among delivery companies, electricity suppliers, and consumers would be helpful to help ensure that consumers will actually see the promised benefits for which they pay.

- Addressing consumer non-financial concerns — a" voice-of-the-consumer" process is needed to provide a mechanism for consumer groups to get their issues and questions presented and addressed. This collaborative process could also serve as a mechanism for collecting good ideas and lessons learned from consumer groups and could improve how the smart grid is ultimately deployed.

Consumer education can provide the understanding, alignment, and motivation to enable active consumer participation on the front end. We must increase our efforts to engage the consumers up front, thus gaining their participation, support, and enthusiasm as we move forward.

The National Energy Technology Laboratory (NETL) continues to support smart grid implementation. Visit the NETL website site to learn more about this effort.

Website: www.netl.doe.gov/smartgrid

Email: smartgrid@netl.doe.gov

BIBLIOGRAPHY

Business Roundtable Report. (2007) "More Diverse, More Domestic, More Efficient: A Vision for America's Energy Future."

Center for Contemporary Conflict Report. (2002) "Economic Costs to the United States Stemming from the 9/11 Attacks."

EPRI. (2001) "The Cost of Power Disturbances to Industrial and Digital Economy Companies."

(2010) "Methodological Approach for Estimating the Benefits and Costs of Smart Grid Demonstration Projects."

(2008) "The Green Grid, Energy Savings, and Carbon Emissions Reductions Enabled by a Smart Grid."

Faruqui, Ahmad. Brattle Group. (2010) "The Ethics of Dynamic Pricing."

Federal Energy Regulatory Committee. (2009) "A National Assessment of Demand Response Potential (FERC staff report). http://www.ferc.gov/legal/staff-reports/06-09-demand-response.pdf.

KEMA. (2008) "The U.S. Smart Grid Revolution: KEMA's Perspective for Job Creation."

Leiby, Paul N., ORNL. (2007) "Estimating the Energy Security Benefits of Reduced U.S. Oil Imports." http://www.epa.gov/oms/renewablefuels/ornl-tm-2007-028.pdf.

Miller, J. (2009) "What's in It for Me? Selling the Smart Grid to Consumers, Smart Grid News." http://www.smartgridnews.com/artman/publish/News_Commentary_News/What-s-in-it-for-Me-Selling-the-Smart-Grid-to-Consumers-1186.html.

Pacific Northwest National Lab. "Impacts Assessment of PHEVs on Electric Utilities and Regional U.S. Power Grids." http://energytech.pnl.gov/publications/pdf/PHEV_Feasibility_Analysis_Part1.pdf.

(2010) "The Smart Grid: An Estimation of the Energy and CO_2 Benefits." http://www.pnl.gov/news/release.aspx?id=776.

Parry, I., M. Walls, and W. Harrington. (2007) "Automobile Externalities and Policies." http://www.rff.org/rff/Documents/RFF-DP-06-26-REV.pdf.

Rand Science and Technology Report. (2004) "Estimating the Benefits of the GridWise Initiative."

Trotter, Joel. (2005) "Safety Programs that Work.". http://www.appanet.org/events/index.cfm?ItemNumber=12577.

U.S. - Canada Power System Outage Task Force. (2004) "Final Report on the August 14th Blackout in the United States and Canada."

U.S. Department of Energy—Advanced Vehicle Testing Activity PHEV Field Test Plans and Testing Results, June 2007, DOE's Advanced Vehicle Testing Activity, PHEV Field Test Plans and Testing Results (PDF)

U.S. Department of Energy, Office of Electricity Delivery and Energy Reliability, National Energy Technology Laboratory. (2009) "Building a Smart Grid Business Case."

http://www.netl.doe.gov/ smartgrid/referenceshelf/ whitepapers/ Whitepaper_Building%20A%20Smart%20Grid%20Business%20Case_APPROVED_2009.pdf.

(2007) "Modern Grid Benefits."

West Virginia Smart Grid Implementation Plan. (2009) http://www.netl.doe.gov/smartgrid/referenceshelf /reports/WV%20SGIP%20Final%20Report%20rev1.pdf.

ACRONYMS

AO	Asset/System Optimization
APPA	American Public Power Association
BAU	Business as usual
CHP	Combined Heat and Power
C&I	Commercial and Industrial
CS	Customer Side Systems
CV	Conventional Vehicle
DER	Distributed Energy Resources
DMS	Distribution Management System/Distribution Automation
DR	Demand Response
EPRI	Electric Power Research Institute
FACTS	Flexible AC Transmission Systems
GDP	Gross Domestic Product
GHG	Greenhouse Gas
GW	Gigawatt
ICT	Information and Communications Integration
IOU	Investor-Owned Utility
IPP	Independent Power Producer
ISO	Independent System Operator
KWH	Kilowatt Hour
LECG	A global expert services firm

LMP	Locational Marginal Price
NO_x	Nitrogen Oxides
NYISO	New York Independent System Operator
O&M	Operating and Maintenance
ORNL	Oak Ridge National Laboratory
PHEV	Plug-in Hybrid Vehicles
PJM	Pennsylvania, New Jersey, Maryland
PNNL	Pacific Northwest National Laboratory
PQ	Power Quality
RAND	A non-profit research and analysis institution
SOA	Service Oriented IT Architecture
SO_x	Sulfur Oxides
TA	Transmission Enhancement Applications
T&D	Transmission and Distribution
T&D O&M	Transmission and Distribution Operations and Maintenance